The Woman who Saved my Life.

Lenny and Marleen were so much in love and they never wanted to part . They stuck together like honey bees. They were the best of lovers. Until one day Lenny met a blonde lady called Triche. He thought he must have been dreaming to see such a lady with such a smooth lean body. Her complexion was out of this world and her hair was as long as Rapunzel's . He thought since he was single nothing should stop him from talking to her. They talked and Triche's speech was as sweet as ever. In the blink of an eye Lenny found himself kissing the lady. "Oh ! What have we just done?" Lenny asked Triche. "Your name is beautiful , who

gave it to you?"Continued Lenny. "Thanks it was my father because he had a lovely sister by that name. "Uttered Triche in a gentle soft voice.

When he got to his home his mobile phone rang and it was Marleen on the other side . She asked him for dinner and he agreed and when he got there Marleen was full of questions. She did not understand why he was not as happy to see her as usual and their kiss was not the same and he also had not showered before the meal as he used to. Lenny said he was in a hurry to come and work was not as fantastic as usual. They had their meal and kissed a bit more and talked of boring stories until they parted for

the night.

The following day Lenny passed through the same place he had met Triche and met her again. This time she was sitted on a bench it was in a park. They greeted each other and kissed. She was eating a burger and offered it to Lenny who refused it because he had just had lunch. They talked and laughed together and played in the park and then left for Lenny's home. He knew Marleen was still at work and would finish a bit late today since it was busy at her work place. Marleen was a doctor and she worked with cancer patients.

When Lenny and Triche got to his home they kissed as they got into the house and they were passionate kisses. They stopped and Lenny caressed her hair which was silky and well maintained. They talked a bit more and drank some wine and had sex with each other. After a few minutes Marleen rang and said she was coming over. When Marleen arrived she could smell some expensive perfume and wondered who had been there.

She was told some lady had just delivered a parcel he had been waiting for and could not say what was in the parcel. Marleen thought maybe she was being played. She walked out on him and told him that maybe

she deserves better. Lenny tried to follow her but she had stopped a taxi and had been driven off. He called her to apologize but she did not pick up her phone. When she got to her place it felt so empty that she cried and called a lady doctor who was Christian like her. She asked her for some advice and firstly she said: "You should not be equally yorked with unbelievers, but I know that Lenny and you have been together for more than five years now. What I would recommend is that you try your best to work things out if not then break up. Maybe he is not really cheating on you but just playing games with you. She told her to read Matthew 6:25. It should help you to calm down and we both had a

busy day so you need some rest.

Marleen tried to sleep but it was difficult then she read the verse she had been given and another one from Galatians 5:19. She just remebered that there were some wine glasses that had just been used at Lenny's place. She then read Ephesians 5:8-11. She then made up her mind to break up with Lenny. She called Lenny and they broke up but he could not believe it he had thought she would beg for him. When she met her friend Clara at work she was very happy and together they celebrated.

That afternoon Marleen met Lenny with

Triche and they greeted her and she went her own way . It still hurt a bit then she called Clara to tell her of what had happened and Clara gave her some other verses from 1Peter 5:7 and 8.She was comforted and went to work the following day as happy and peaceful as ever.

Marleen then got a secret admirer from work who was also a believer. They fell in love and went to the same church on Sunday. Clara who was now planning to get married to a gynaecologist was very pleased for her. Marleen had fallen in love with an opthamologist.They liked the fact that they were all Christians.She was learning to be an opthamologist too.

Attending class was enjoyable for Marleen and the opthamologist who sometimes taught her in some of their classes. They had already fallen in love before she made arrangements to attend his classes. The classes were super and Marleen was a first class student. After classes Darren Mcclaren her tutor and husband to be would drive her home. She made sure she left her ford fiesta at home on days she was going to meet him. Darren and Marleen were so much in love that they wanted to marry each other.

Clara and Frank the gynaecologist married

and had a baby girl called Tamara. Clara had a baby welcome party for her baby and she received many presents for her baby that came from Mother Care. She was very happy for Tamara. Darren and Marleen where about to get married too. They invited all the doctors from Northern General Hospital. The doctors had come with their partners. There were a few relatives who had been invited it was a beautiful wedding with speeches and dancing as well as kissing from different lovers and from the bride and bridegroom. Music from different gospel and country singers was played as the cutting of the cake was in process. After people had had the cake .There was the throwing of the

boquet part and it was a lady from Marleen's family who caught it . Her name was Felistas and she had a boyfriend whom she had come with to the wedding.

They were given plenty of presents and Darren's mum, Jess gave the vote of thanks .She thanked all the doctors that had come and gave some of them flowers and the others chocolates. She said that it was wonderful to see all the relatives that came and they were given sweets and chocolate biscuits. The day had been fabulous and fun filled.

Marleen passed her exams in opthamology and was now well qualified to open a surgery together with her husband Dr Macclaren.They had a baby boy whom they

called Jimmy. Dr Marleen and Dr Darren bought a new home for themselves in Dore in Sheffield. They lived happily afterwards together and used to have a number of visitors.

The Woman who Saved my Life.

Lenny and Marleen were so much in love and they never wanted to part . They stuck together like honey bees. They were the best of lovers. Until one day Lenny met a blonde lady called Triche. He thought he must have been dreaming to see such a lady with such a smooth lean body. Her complexion was out of this world and her hair was as long as Rapunzel's . He thought since he was single nothing should stop him from talking to her. They talked and Triche's speech was as sweet as ever. In the blink of an eye Lenny found himself kissing the lady. "Oh ! What have we just done?" Lenny asked Triche. "Your name is beautiful , who

gave it to you?"Continued Lenny. "Thanks it was my father because he had a lovely sister by that name. "Uttered Triche in a gentle soft voice.

When he got to his home his mobile phone rang and it was Marleen on the other side . She asked him for dinner and he agreed and when he got there Marleen was full of questions. She did not understand why he was not as happy to see her as usual and their kiss was not the same and he also had not showered before the meal as he used to. Lenny said he was in a hurry to come and work was not as fantastic as usual. They had their meal and kissed a bit more and talked of boring stories until they parted for

the night.

The following day Lenny passed through the same place he had met Triche and met her again. This time she was sitted on a bench it was in a park. They greeted each other and kissed. She was eating a burger and offered it to Lenny who refused it because he had just had lunch. They talked and laughed together and played in the park and then left for Lenny's home. He knew Marleen was still at work and would finish a bit late today since it was busy at her work place. Marleen was a doctor and she worked with cancer patients.

When Lenny and Triche got to his home they kissed as they got into the house and they were passionate kisses. They stopped and Lenny caressed her hair which was silky and well maintained. They talked a bit more and drank some wine and had sex with each other. After a few minutes Marleen rang and said she was coming over. When Marleen arrived she could smell some expensive perfume and wondered who had been there.

She was told some lady had just delivered a parcel he had been waiting for and could not say what was in the parcel. Marleen thought maybe she was being played. She walked out on him and told him that maybe

she deserves better. Lenny tried to follow her but she had stopped a taxi and had been driven off. He called her to apologize but she did not pick up her phone. When she got to her place it felt so empty that she cried and called a lady doctor who was Christian like her. She asked her for some advice and firstly she said: "You should not be equally yorked with unbelievers, but I know that Lenny and you have been together for more than five years now. What I would recommend is that you try your best to work things out if not then break up. Maybe he is not really cheating on you but just playing games with you. She told her to read Matthew 6:25. It should help you to calm down and we both had a

busy day so you need some rest.

Marleen tried to sleep but it was difficult then she read the verse she had been given and another one from Galatians 5:19. She just remebered that there were some wine glasses that had just been used at Lenny's place. She then read Ephesians 5:8-11. She then made up her mind to break up with Lenny. She called Lenny and they broke up but he could not believe it he had thought she would beg for him. When she met her friend Clara at work she was very happy and together they celebrated.

That afternoon Marleen met Lenny with

Triche and they greeted her and she went her own way. It still hurt a bit then she called Clara to tell her of what had happened and Clara gave her some other verses from 1Peter 5:7 and 8. She was comforted and went to work the following day as happy and peaceful as ever.

Marleen then got a secret admirer from work who was also a believer. They fell in love and went to the same church on Sunday. Clara who was now planning to get married to a gynaecologist was very pleased for her. Marleen had fallen in love with an opthamologist. They liked the fact that they were all Christians. She was learning to be an opthamologist too.

Attending class was enjoyable for Marleen and the opthamologist who sometimes taught her in some of their classes. They had already fallen in love before she made arrangements to attend his classes. The classes were super and Marleen was a first class student. After classes Darren Mcclaren her tutor and husband to be would drive her home. She made sure she left her ford fiesta at home on days she was going to meet him. Darren and Marleen were so much in love that they wanted to marry each other.

Clara and Frank the gynaecologist married

and had a baby girl called Tamara. Clara had a baby welcome party for her baby and she received many presents for her baby that came from Mother Care. She was very happy for Tamara. Darren and Marleen where about to get married too. They invited all the doctors from Northern General Hospital.   The doctors had come with their partners. There were a few relatives who had been invited it was a beautiful wedding with speeches and dancing as well as kissing from different lovers and from the bride and bridegroom. Music from different gospel and country singers was played as the cutting of the cake was in process. After people had had the cake .There was the throwing of the

boquet part and it was a lady from Marleen's family who caught it . Her name was Felistas and she had a boyfriend whom she had come with to the wedding.

They were given plenty of presents and Darren's mum, Jess gave the vote of thanks .She thanked all the doctors that had come and gave some of them flowers and the others chocolates. She said that it was wonderful to see all the relatives that came and they were given sweets and chocolate biscuits. The day had been fabulous and fun filled.

Marleen passed her exams in opthamology and was now well qualified to open a surgery together with her husband Dr Macclaren.They had a baby boy whom they

called Jimmy. Dr Marleen and Dr Darren bought a new home for themselves in Dore in Sheffield. They lived happily afterwards together and used to have a number of visitors.

The Woman who Saved my Life.

Lenny and Marleen were so much in love and they never wanted to part . They stuck together like honey bees. They were the best of lovers. Until one day Lenny met a blonde lady called Triche. He thought he must have been dreaming to see such a lady with such a smooth lean body. Her complexion was out of this world and her hair was as long as Rapunzel's . He thought since he was single nothing should stop him from talking to her. They talked and Triche's speech was as sweet as ever. In the blink of an eye Lenny found himself kissing the lady. "Oh ! What have we just done?" Lenny asked Triche. "Your name is beautiful , who

gave it to you?"Continued Lenny. "Thanks it was my father because he had a lovely sister by that name. "Uttered Triche in a gentle soft voice.

When he got to his home his mobile phone rang and it was Marleen on the other side . She asked him for dinner and he agreed and when he got there Marleen was full of questions. She did not understand why he was not as happy to see her as usual and their kiss was not the same and he also had not showered before the meal as he used to. Lenny said he was in a hurry to come and work was not as fantastic as usual. They had their meal and kissed a bit more and talked of boring stories until they parted for

the night.

The following day Lenny passed through the same place he had met Triche and met her again. This time she was sitted on a bench it was in a park. They greeted each other and kissed. She was eating a burger and offered it to Lenny who refused it because he had just had lunch. They talked and laughed together and played in the park and then left for Lenny's home. He knew Marleen was still at work and would finish a bit late today since it was busy at her work place. Marleen was a doctor and she worked with cancer patients.

When Lenny and Triche got to his home they kissed as they got into the house and they were passionate kisses. They stopped and Lenny caressed her hair which was silky and well maintained. They talked a bit more and drank some wine and had sex with each other. After a few minutes Marleen rang and said she was coming over. When Marleen arrived she could smell some expensive perfume and wondered who had been there.

She was told some lady had just delivered a parcel he had been waiting for and could not say what was in the parcel. Marleen thought maybe she was being played. She walked out on him and told him that maybe

she deserves better. Lenny tried to follow her but she had stopped a taxi and had been driven off. He called her to apologize but she did not pick up her phone. When she got to her place it felt so empty that she cried and called a lady doctor who was Christian like her. She asked her for some advice and firstly she said: "You should not be equally yorked with unbelievers, but I know that Lenny and you have been together for more than five years now.What I would recommend is that you try your best to work things out if not then break up. Maybe he is not really cheating on you but just playing games with you. She told her to read Matthew 6:25. It should help you to calm down and we both had a

busy day so you need some rest.

Marleen tried to sleep but it was difficult then she read the verse she had been given and another one from Galatians 5:19. She just remebered that there were some wine glasses that had just been used at Lenny's place. She then read Ephesians 5:8-11. She then made up her mind to break up with Lenny. She called Lenny and they broke up but he could not believe it he had thought she would beg for him. When she met her friend Clara at work she was very happy and together they celebrated.

That afternoon Marleen met Lenny with

Triche and they greeted her and she went her own way . It still hurt a bit then she called Clara to tell her of what had happened and Clara gave her some other verses from 1Peter 5:7 and 8.She was comforted and went to work the following day as happy and peaceful as ever.

Marleen then got a secret admirer from work who was also a believer. They fell in love and went to the same church on Sunday. Clara who was now planning to get married to a gynaecologist was very pleased for her. Marleen had fallen in love with an opthamologist.They liked the fact that they were all Christians.She was learning to be an opthamologist too.

Attending class was enjoyable for Marleen and the opthamologist who sometimes taught her in some of their classes. They had already fallen in love before she made arrangements to attend his classes. The classes were super and Marleen was a first class student. After classes Darren Mcclaren her tutor and husband to be would drive her home. She made sure she left her ford fiesta at home on days she was going to meet him. Darren and Marleen were so much in love that they wanted to marry each other.

Clara and Frank the gynaecologist married

and had a baby girl called Tamara. Clara had a baby welcome party for her baby and she received many presents for her baby that came from Mother Care. She was very happy for Tamara. Darren and Marleen where about to get married too. They invited all the doctors from Northern General Hospital. The doctors had come with their partners. There were a few relatives who had been invited it was a beautiful wedding with speeches and dancing as well as kissing from different lovers and from the bride and bridegroom. Music from different gospel and country singers was played as the cutting of the cake was in process. After people had had the cake .There was the throwing of the

boquet part and it was a lady from Marleen's family who caught it . Her name was Felistas and she had a boyfriend whom she had come with to the wedding.

They were given plenty of presents and Darren's mum, Jess gave the vote of thanks .She thanked all the doctors that had come and gave some of them flowers and the others chocolates. She said that it was wonderful to see all the relatives that came and they were given sweets and chocolate biscuits. The day had been fabulous and fun filled.

Marleen passed her exams in opthamology and was now well qualified to open a surgery together with her husband Dr Macclaren.They had a baby boy whom they

called Jimmy. Dr Marleen and Dr Darren bought a new home for themselves in Dore in Sheffield. They lived happily afterwards together and used to have a number of visitors.

www.ingramcontent.com/pod-product-compliance
Lightning Source LLC
Chambersburg PA
CBHW021857170526
45157CB00006B/2495